Nature Watch: Animal Homes

CONTENTS

Animal homes	2
Warm & shady woodland homes	4
Leaves — wanted dead or alive!	6
Who's afraid of the dark?	8
Green & leafy rainforest homes	10
Wet wonderland	12
Ground floor, going up!	14
Hot & dry desert homes	16
Water in the desert	18
When rain falls in the desert	20
Watery & wild wetland homes	22
Living on the edge & in the deep	24
Banking on it!	26
Salty & sunny seaside homes	28
The tides are turning!	30
In the shallows & the deep blue sea	32
Fancy & fragile coral reef homes	34
Colourful coral	36
My island home	38
Wild animals at home in the suburbs	40
Around your house & in the neighbourhood	42
Animals in your garden	44
Attracting animals to your backyard	46
Glossary	48

INTRODUCTION

Animal homes

Koala

All living things need somewhere to live and almost every space on Earth has someone or something living in it. Animals and plants have different shapes, colours and ways of life to help them survive. There are as many different kinds of animals as there are habitats or places to live. It doesn't matter whether it is the driest desert or the deepest ocean, some kind of living thing will have made it a home.

Different kinds of homes

Woodlands, rainforests, rivers, ponds, oceans, seashores and many other places provide homes for thousands of animals. You might even find some animals living with you in your very own house!

Possum

Baby birds

WOODLAND HOMES

Warm & shady woodland homes

Kookaburra

Barking owl

Places that are mostly covered by trees are called woodlands. Animals can find lots of hiding spots in woodlands. They may hide in the tops of the trees, under bark or in the grass and leaf litter on the ground.

Possums live high in the tree tops where they search for flowers, fruit and leaves to eat.

Geckoes and other lizards scurry about on bark or among shrubs, looking for insects.

Kookaburras hunt for animals crawling on the ground during the day, and owls hunt these same animals at night.

Brushtail possums

It's the small things that matter

Many flying and crawling insects carry pollen from flower to flower in the woodland. This helps the flowers make seeds so new trees and other plants can grow. Without insects, woodlands would look very different to the way they do today.

Butterfly
Beetle
Ant

I love my home among the gum trees.

Wombat

Echidna

WOODLAND HOMES

Gecko

Leaves – wanted dead or alive!

Dead or alive, the woodland's leaves give life to the animals. When fresh and green, they provide oxygen, food and shelter to animals like koalas, possums and birds. Even when the leaves have died and fallen to the ground, they still provide homes and food.

Black snakes may make their homes among dead leaves on the ground. They slide silently through the leaf litter, looking for frogs to catch and eat.

Button-quails and other small birds live on the ground. They have brown speckled feathers that help them to camouflage or blend into the fallen leaves and grass.

Red-bellied black snake

Button-quail

Pademelons

Falling food

As well as leaves, fruits fall onto the ground where they may be found and eaten by the animals that live on the forest floor.

Koala

Magpies

Life at the top

Many birds build their nests high in the tree tops where the chicks will be out of reach of ground animals. Koalas also live in trees to stay safe from predators. They eat only gum leaves and usually don't drink water, so they don't need to come down very often.

WOODLAND HOMES

Brushtail possum

Who's afraid of the dark?

Spider

Some of the most important homes to woodland animals are hollows in trees, burrows in the ground and rocky caves. These dark spaces are perfect for keeping out the wind and rain, and for hiding from predators.

Wallaby

Some places may be large enough for many different kinds of animals to share. In caves, there may be bats hanging from the ceiling, spiders and insects crawling on the walls, and wallabies, kangaroos and other animals scurrying around on the ground.

No need to see

Bats use "echolocation" to help them fly around in the dark without bumping into each other! They send out sounds and listen to the echoes that bounce off the surrounding objects.

Numbat

Bats

Glider

Hollow homes

Bats, possums, gliders, mice, parrots and even numbats shelter in hollows. Tree hollows take a long time to form, so humans shouldn't cut down old or dead trees because they are special homes for many animals.

RAINFOREST HOMES

Butterfly

Cassowary

Green & leafy rainforest homes

Rainforests are places where the forest grows so thick and lush that the tree tops join to form a "canopy" and their leaves block out most of the sunshine. The leafy canopy acts a little like an umbrella except it keeps the moisture inside the forest.

Leaf and fruit eaters like possums and tree-kangaroos make their homes in rainforests and are surrounded by food. But life isn't always easy. They have to be careful not to get caught and eaten by predators. Snakes, especially pythons, are always on the hunt for a tasty meal like a possum or tree-kangaroo.

Green possum

Tree-frog

Not so hard to swallow

Lush, wet rainforest is the perfect home for insects of all shapes and sizes. Frogs enjoy these moist surroundings too. They also enjoy the insects, which they regularly swallow whole.

I love to eat my greens!

Tree-kangaroo

Green tree python

RAINFOREST HOMES

Freshwater catfish

Crayfish

Wet wonderland

Turtle

Cool, clear freshwater pools and streams are often found in rainforests. All kinds of animals live in them, including tiny insects, crayfish, leeches, fish, frogs, turtles and even some mammals, such as the platypus.

Turtles love climbing onto logs and stones at the water's edge to soak up any sunshine that reaches the forest floor. Turtles are well suited for living in rainforest streams. They have webbed feet for swimming as well as claws for climbing.

Many fish in rainforest ponds feed on insects that fall into the water. Some can even shoot the insects with a squirt of water.

Archerfish

Hunting with your eyes closed

Clear, fast-flowing rainforest streams are excellent homes for platypuses. These furry animals use their sense of touch to find their food underwater.

Platypus

Female (top) and male (bottom) eclectus parrots

Striped possum

Tree-frog

Water views can be dangerous

Rainforest streams can be dangerous. Many animals risk drowning or being eaten by fish or eels if they fall in the water. Possums and gliders avoid the water and climb or glide from tree to tree to cross rainforest streams.

RAINFOREST HOMES

Ground floor, going up!

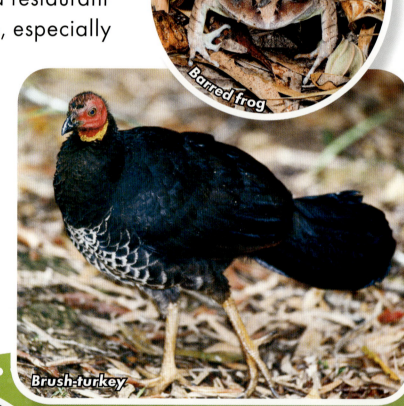

Green tree python

The sun rarely shines on the floor of the rainforest. Many animals make their homes down here in the shadows. Some eat leaves, seeds and fruit that fall from the trees. These animals might later be eaten by predators like quolls or pythons. The forest floor is like a restaurant and it can be a very busy place, especially at night.

Brush-turkeys are ground birds and rarely fly far. They spend their time turning over leaves, looking for insects, worms and snails that live on the damp forest floor.

Many frogs enjoy the damp conditions on the ground. Most frogs have skin colours that match their surroundings and make them hard for predators to find.

Barred frog

Brush-turkey

Bandicoot

Keeping dry

Some animals, like this bandicoot, tunnel under the leaf litter or pile it up over themselves to hide and to keep dry when it rains.

Spotted-tailed quolls

Getting off the ground

Many birds don't like to be close to the ground where predators can catch them. They nest in tree hollows well above the forest floor, but they aren't always safe there. Pythons inhabit both the forest floor and the tree tops and are excellent climbers that can easily find them.

Carpet Python

DESERT HOMES

Bilby

Hot & dry desert homes

Nailtail wallaby

Much of Australia is desert. In some places, the dry red sand stretches out for hundreds of kilometres and may not receive rain for many years. The animals living here have to be tough to survive.

Sand monitor

Wallabies and kangaroos that live in these dry places stay out of the daytime heat. They only come out to feed early in the morning or late in the afternoon.

Emus are very hardy birds, but if they come out in the daytime they have to pant to keep cool.

Emus

Can you dig it?

Many desert animals, including snakes like the woma python, burrow underground to get away from the hot sun. The bilby also stays underground during the day. It digs its burrows near spinifex bushes where the soil is firmer.

Woma

DESERT HOMES

Water in the desert

Dragonfly

Mulgara

Even in a dry country, animals need water to live. Some animals get all of their water from the food they eat and never have to drink. Most, however, do need to drink and when they find water in the desert they don't wander too far away from it.

Both the mulgara and the spinifex hopping-mouse get the water they need from their food. The mulgara eats insects and lizards, while the spinifex hopping-mouse mainly eats seeds and plants.

The galah can live in some of the driest parts of Australia, but it still needs to find water to drink. The number of galahs has increased in areas where people have settled and built windmills and water bores.

Galah

Spinifex hopping-mouse

Snapping-frog

Underground frogs

Not all frogs need to live in ponds! Some frogs make their homes in the desert and burrow into the ground when the weather is very dry. They may stay underground for a very long time and only dig themselves out when it rains.

Sometimes in the morning, water droplets called dew form on the ground. Dew may also collect on the thorny devil's back. Its spiny skin makes the dew run towards its mouth so it can drink.

Thorny devil

Golden perch

Waiting for rain

Fish like the golden perch can live in muddy desert streams and pools. They won't lay any eggs until it rains and the water level in their creek or pond begins to rise.

DESERT HOMES

Crimson chat

Wedge-tailed eagle

When rain falls in the desert

In places where just enough rain falls each year, a few trees and some grass may grow. For most of the year it is very dry, but the trees and shrubs find enough water to keep themselves, and lots of animals, alive.

After rain, the grass and trees turn green and grow quickly, which means there is more food for insects and birds like the crimson chat.

Insect numbers boom and so do insect-eating animals like the earless dragon. This then causes the number of meat-eating predators to increase. More food always means more animals. Desert creatures have to make the most of this food-filled time because the hot sun quickly dries up the water and the trees and grass die out again.

Earless dragon

All you can eat!

Because of the increase of food after it rains in the desert, the number of rats, mice and other small mammals grows. Many of these animals are caught and eaten by eagles, snakes and other predators.

Death adder

Wallaroos and rock-wallabies are steady boulder-hopping marsupials. They find water where it collects after rain in the spaces between rocks.

Rock-wallaby

Wallaroo

Spinifex pigeon

A need for seed

The spinifex pigeon never wanders too far from water and makes itself at home in rocky outcrops. When it rains, the grasses grow and produce seeds that the spinifex pigeon needs for food.

WETLAND HOMES

Watery & wild wetland homes

Barramundi

Black-necked stork

Freshwater lagoons, swamps and marshes have all kinds of homes for animals. There are homes above the water, below the surface, on the banks, in the deep and in the shallows.

Every animal has special features to suit its home. Swans have webbed feet for paddling in deep water. Their long necks help them reach the bottom to look for food. Storks have long legs to wade through shallow water without getting their feathers wet. Their thick, powerful bills are useful for catching fish.

Black swans

Gudgeon

Under the surface

Turtles and fish live under the water, although turtles need to come up to breathe. Crocodiles spend a great deal of time floating at the surface, waiting for their next meal.

We love having a dip!

Estuarine crocodiles

Long-necked turtle

WETLAND HOMES

Living on the edge & in the deep

Shrimp

Dwarf tree-frog

At the water's edge, the reeds and lilies grow thickest. Their roots grip the muddy bottom in the shallow water. Insects, snails, shrimps and lots of small animals live here, feeding on the plants and on each other.

Frogs search for insects among the reeds. Many frogs are exactly the same colour as the leaves they sit on, which helps them to catch their prey and to hide from predators.

Some snakes, such as the water python, are very good swimmers. Like other pythons, they catch and squeeze their prey to death before swallowing it whole.

Water python

Pain in the neck!

White-necked heron

This heron looks like it has a broken neck but it is perfectly fine. The bend in the neck allows the heron to pull its head right back then swiftly strike forward and spear fish in the water.

The very rare freshwater sawfish uses its long spiky nose to dig freshwater mussels and crayfish out of the mud on the river bottom.

Freshwater sawfish

In deep water

Many fish live in murky freshwater rivers. The biggest freshwater fish in Australia is the Murray cod. It may grow larger than an adult person – and can weigh twice as much! Sadly, the Murray cod is vulnerable because humans have built dams and taken away too much water from its river home.

Murray cod

WETLAND HOMES

Honeyeater

Osprey

Banking on it!

While many animals live in the water, even more live on the banks that surround it. These wetland dwellers and visitors come down to the water to drink or hunt for food.

Ospreys build their nests in tall trees that grow on the banks. Every day they go fishing, flying high over the water, keeping a watchful eye on the surface for ripples that tell them where the fish are swimming.

Crocodiles

Crocodiles and turtles spend a long time out of the water sunning themselves on the banks. They need to warm their bodies so they can move quickly to catch food.

Sitting on the sideline

Pelicans and other waterbirds sometimes have to share fallen logs with each other when drying out their feathers.

Logs that have been soaking in water are also excellent homes for worms and burrowing insects.

Pelicans

Dingo

Wallaroo

Just Visiting

Wallaroos, like many animals, come down to the water to drink in the early morning or late afternoon. While drinking, they stay alert and twitch their ears to listen for danger. Their enemy the dingo drinks here too.

SEASHORE HOMES

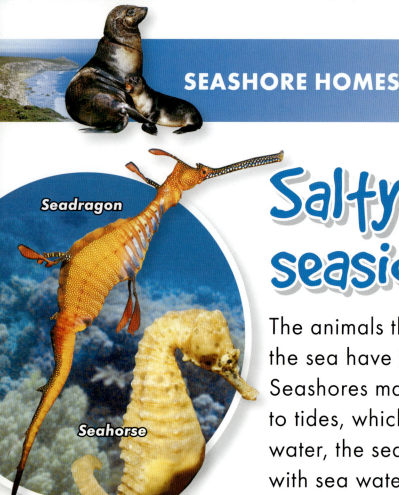

Seadragon

Seahorse

Penguins

Salty & sunny seaside homes

The animals that live where the land meets the sea have to survive in a wet, salty world. Seashores may be sandy, rocky or muddy. Due to tides, which are the rise and fall of the ocean's water, the seashore may be completely covered with sea water (high tide) or exposed to the hot sun (low tide).

Fur-seal

Animals that live on sandy beaches are very different in shape and size to those that swim in the sea, live among the mangroves or cling to the rocks. Each seaside home has its own special group of animals.

Drifters & suckers

Box jelly

Sea animals have learned to move around in different ways. The box jelly uses a pumping action to move as fast as an Olympic swimmer! Others, like some sea stars, use suckers to grip onto and crawl along the sea floor. Some animals drift around and go wherever the ocean takes them!

Jump in, the water is fine!

Fish & coral

Sea star

SEASHORE HOMES

The tides are turning!

Octopus

Tides play an important part in the lives of beach animals. At low tide, sand or rocks are uncovered. Some animals use this time to lie in the sun and warm up, while those that need water, hide in the shallow pools left behind after high tide.

In southern Australia, sea-lions pull themselves out of the water at low tide and find a spot on the beach to sunbake and make friends.

Terns spend much of their time fishing at sea. If the time of day and tides are not suitable for fishing, they settle in a safe place and go to sleep facing the wind.

Sea-lions

Terns

Stingray

When the tide comes in

Rays and other fish move into the shallows on the incoming tide, which is when the water is rising to reach high tide. As the sea covers the beach once more, the fish search for food.

Pineapplefish

Crayfish

Sea star

Anemone

Don't get washed away

In the shallows and rock pools formed by the tides, anemones, sea stars, crabs and small fish find shelter. Many of the animals living here have to cling on tightly to the rocks so they don't get washed off by the surf when the tide comes back in.

SEASHORE HOMES

In the shallows & the deep blue sea

Crab

Blue-ringed octopus

Many small fish live in the warm, shallow waters of mudflats and mangroves. Sunlight can easily reach the bottom, where plants grow and feed the fish. Occasionally, much larger animals from more open waters of the sea come in to the shallows to hunt smaller fish.

Pied stilts

Crab

Thousands of birds also hunt in shallow waters, where they dig up small animals to eat. Tiny crabs living in the mud have hard shells and big claws for protection but some birds swallow them whole — claws, shell and all!

Fish out of water

Mudskipper

Mudskippers are small fish that can spend a lot of time out of water. They live in mudflats and have strong front fins to help them crawl over the mud at low tide.

White sharks travel the open ocean but hunt seals and other mammals that live close to the shore. They are very powerful and fast because they can keep their body temperature higher than the temperature of the surrounding water.

White shark

Helpful hunters

Dolphin

Dolphins often hunt small fish close to shore. Aborigines used to look for dolphins so they knew where the fish were gathering. Sometimes the dolphins would work with the Aborigines to help them catch the fish.

CORAL HOMES

Murex shell

Fancy & Fragile coral reef homes

Lionfish

Cowrie

Corals are not rocks or plants, but animals! Each tiny coral is a soft-bodied animal called a polyp. Millions of these polyps grow together and form hard, stony skeletons on the outside of their bodies. Over thousands of years, the coral skeletons build up on top of each other until they form a reef.

Coral reefs are excellent homes for thousands of animals, including fish, sea stars, sea cucumbers and turtles. Because so many animals live in reefs, it is very important that humans take care of these precious homes.

Sea star & coral

In the tube

Many other animals that live in reefs also have their skeletons on the outside. This Christmas tree worm lives inside a red tube and sticks up two "crowns" covered in little tentacles to catch food.

Christmas tree worm

Our backyards are very pretty.

Long-nosed butterflyfish

Clownfish & anemone

CORAL HOMES

Sea whip

Sea snake

Colourful coral

Living coral can be brightly coloured and is often named after its shape. Fan coral is shaped like a fan. Mushroom coral looks like a mushroom and you can guess what brain coral looks like! People often go snorkelling over reefs to look at the beautiful corals and the animals that live among them.

Nudibranch

Queensland's Great Barrier Reef is the largest reef in the world. We must look after the Great Barrier Reef because corals are very sensitive and even the smallest change in water temperature can kill them. Corals turn white when they die, which is called "coral bleaching", and other sea animals cannot live in dead coral.

Anemonefish

Coral polyps

A polyp if you please

Corals, just like all other animals, need to eat. Each coral polyp has a mouth and can catch tiny food with its tentacles. But special algae that live on the coral also give it energy, which the algae gets from the sun.

Coral

Feather star

Greyface moray eels

Green moray eel

Hiding holes

Coral can form underwater caves and hollows, which are ideal hiding places for animals like these moray eels. Moray eels have lots of sharp teeth and ambush their prey.

CORAL HOMES

My island home

Turtles

Coral islands form when waves break off bits of coral from the reef and pile the pieces together. Then seeds, like coconuts, get washed up onto shore and grow into trees. Birds also drop seeds as they fly over the island. Insects and small reptiles get carried to the island by the wind or on driftwood. Soon the coral island becomes home to hundreds of animals.

Terns

Butterfly

Masked booby

Sea birds use coral islands as places to rest. Some birds build their nests on the beaches or in the trees that grow on the island. They spend most of their days fishing at sea.

Turtle hatching

Green turtle

Buried treasure

On special nights of the year, marine turtles drag themselves out of the sea and up the beaches of chosen coral islands. They dig deep holes and lay eggs into them. The turtles return to sea before the eggs hatch.

Coral island lagoon

Sea star

Stingray

Giant clam

Lovely lagoons

Many coral islands have lagoons. Sea water in these lagoons is protected from the waves and the wind by the island. Lots of animals, such as sea stars, make their home in the calm waters. Sometimes coral surrounds the lagoon and may show above the water at low tide.

OUR HOMES

Wild animals at home in the suburbs

Spider

Possum

People build houses to live in, but they aren't the only ones living in them. You have probably been sharing your house with several animals and you may not have known. Usually they find a place where the owners don't notice them.

Possums, birds, lizards, frogs, spiders and insects have all lived alongside us. The houses we build are much better at keeping out the wind and rain than a nest or hole in the ground.

Blue-tongue Lizard

Skink

Sitting on the fence

Skinks and other lizards love to sun themselves on fence posts, front steps and garden paths. They also love to eat the insects and snails that visit the plants in our garden.

Your home is our home!

Barn owls

Rosella

OUR HOMES

Around your house & in the neighbourhood

Skink

Carpet snake

Our towns and cities may not be the best places for all animals to live but a few clever ones have learned how to live successfully with people. Our parks and buildings give them shelter and sometimes keep them safe from predators.

Rats and mice can live in our houses and steal our food. Carpet snakes may be attracted to our homes by the rats and mice. Many farmers even keep carpet snakes in the roof to keep the number of mice down.

Birds feel safe on electric powerlines high above the ground where they can keep a look out for danger.

Galahs

A light meal

Moths are attracted to house lights at night. Geckoes live under window ledges and dart out to catch the moths. Geckoes have sticky pads on their feet, which allow them to run upside-down across ceilings to catch food.

Moth

Gecko

When native habitats are cleared around our towns and cities animals like flying-foxes may have to move into our parklands to live.

Flying-foxes

Brushtail possum

Kept in the dark

Possums love dark, dry spaces inside the roofs of buildings and may build their nests, called dreys, there. Many people don't even know they have possums living in their ceilings.

OUR HOMES

Ladybirds

Animals in your garden

People plant trees and put objects in their gardens that animals can use for their own purposes. Something like a garden shed makes an excellent place to shelter. A leaking tap can be perfect for a bird to get a quick drink of water.

Honeyeater

Some people love watching birds so they place bird baths and feeding stations in the garden. Birds enjoy having a wash in bird baths and visit the feeding stations every day.

Honeyeaters

Birds like wagtails and fantails hunt insects that visit the flowers and other plants in the garden. They may even nest in the trees and shrubs in the backyard.

Fantail

As still as a log

Frogmouths are hard to see in your backyard. Their feathers are the same colour as tree bark and they sit so still that they look like a branch or piece of wood.

Frogmouths

Toes instead of tongues!

Flowers in the garden attract birds and insects. A butterfly has tastebuds on the bottom of its feet. When it stands on a flower, it is tasting the flower with its toes!

OUR HOMES

Attracting animals to your backyard

Rosella

Animals will visit gardens where they can find food and shelter and where they feel safe. Unfortunately, cats and dogs frighten native animals away. More animals visit gardens that don't have cats or dogs.

The food we eat is not always good for animals so we should not give them our food. It's best to grow plants that the animals usually eat in the wild.

Ringtail possum

Rosella

Building an animal home

Nest boxes are good animal homes. You can build them with different sized openings to suit a variety of animals. Look online for ideas and designs for nest boxes to build with Dad or Mum.

Make your garden safe for me!

Sugar glider

Bearded dragon

Lorikeet

How do I get animals to visit?

Ask Mum or Dad about the best native trees to plant. You can also place hollow logs in the garden to make homes for gliders, possums, bats and birds.

GLOSSARY

ALGAE A type of plant.

AMBUSH Hiding to attack by surprise.

ATTRACT To draw in; cause something to approach.

AVOID To keep away from.

CAMOUFLAGE To blend into the background using colours and shapes.

CANOPY The leafy branches of trees that form a thick cover over the plants underneath.

DREY The nest of a possum or glider.

DRIFTWOOD Wood that floats on, or gets washed up onshore by, water.

ECHOLOCATION To sense an object by sending out sounds then listening to the echoes that bounce back off the object.

EXPOSED Bare or without protection.

HABITAT The place where an animal or plant lives or grows.

HIGH TIDE When the ocean's water level is at its highest point for the day.

INHABIT To live in.

LEAF LITTER Dead leaves and other matter from a tree that has fallen onto the ground.

LOW TIDE When the ocean's water level is at its lowest point for the day.

MAMMAL A class of animals that all have hair on their bodies and feed their babies milk.

MARINE Relating to the sea.

NATIVE Belonging to a particular region or country; not introduced.

POLLEN The yellow powdery grains from a flower that are needed to produce more flowers.

PREDATOR An animal that hunts and eats animals.

RARE Not common.

VULNERABLE Not protected.